Astronomy

Graham Peacock and Dennis Ashton

Wayland

Titles in the series:

ASTRONOMY • ELECTRICITY • FORCES
GEOLOGY • HEAT • LIGHT • MATERIALS
METEOROLOGY • SOUND • WATER

Editor: Polly Goodman
Series Designer: Jan Sterling, Sterling Associates
Book Designer: Joyce Chester
Consultant: Jane Battell, Science Advisory Teacher
Photo Stylist: Zoë Hargreaves

First published in 1994 by
Wayland (Publishers) Ltd
61 Western Road, Hove
East Sussex, BN3 1JD, England

© Copyright 1994 Wayland (Publishers) Ltd

British Library Cataloguing in Publication Data

Peacock, Graham
Astronomy.–(Science Activities Series)
I. Title II. Ashton, Dennis III. Series
522

ISBN 0 7502 1257 8

Acknowledgements
The publishers would like to thank the following for allowing their pictures to be used in this book:
Bryan & Cherry Alexander p11; Chapel Studios p5; Image Select p20 (bottom left); Paul Crowder & Steve Ibbotson p23 (top right), 24 (left & right), 27 (bottom left); Science Photo Library *Cover* (centre & bottom right), p7, 13, 14, 20 (bottom right), 23 (bottom right), 25, 27 (top right), 28. All commissioned photographs are from the Wayland Picture Library (Zul Mukhida). All artwork is by Tony de Saulles.

The publishers would also like to thank the pupils, parents and teachers of Somerhill Road Junior School, Hove, for their help in making this book.

Typeset by Joyce Chester Typographics
Printed and bound in Italy by G. Canale & C.S.p.A.

Contents

The turning Earth	4
Day and night	6
Seasons	8
Seasonal changes	10
The sky in daylight	12
The sky at night	13
The Moon's surface	14
Phases of the Moon	16
Planet models	18
Observing planets	20
Constellations	22
Observing the night sky: March–August	24
Observing the night sky: September–February	26
Meteors	28
Glossary	30
Books to read	31
Notes for parents and teachers	31
Index	32

Words that appear in **bold** are explained in the glossary on page 30.

The turning Earth

We live on a **planet** which is whirling through space. As it spins it gives us day and night. It is tilted on its **axis** and circling the Sun, which gives us **seasons**. There are eight other planets **orbiting** the Sun, but it seems that the Earth is the only one on which there is life.

The stars are so far away that their light has spent thousands of years travelling to us. In this book you will explore the Earth's place in the universe. You will discover ways to look at the night sky and some of the wonders which it contains. This is the study of astronomy.

How do sundials work?

You will need:
- a sunny place outside
- a thin cane
- a big sheet of paper
- a felt tip pen
- a plastic cup
- 4 weights

1. Push the cane into the ground in a sunny place. If the ground is too hard, push the cane into a container of sand or soil.

2. Rest the plastic cup upside down on top of the cane to protect your eyes from the cane.

3. Put weights on the corners of the paper.

4. Draw around the shadow position on the paper every hour and write down the time beside it.

If you were to look down at the North **Pole** from space, you would see the Earth rotating anticlockwise. From the Earth, this rotation makes it look as if the Sun moves from east to west across the sky.

Make a shadowmeter

You will need:
- a sunny place
- a ruler
- a pencil
- a sheet of A4 paper

1 Using the measurements shown in the diagram below, draw nine lines across the sheet of paper.

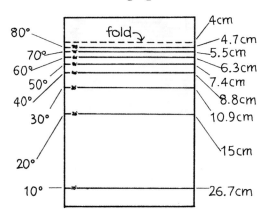

2 Label the lines with the angles.

3 Make a crease along the first 4 cm line and fold the paper up at an angle of 90°.

4 Put the shadowmeter on a table in the sunlight so that the folded end makes a shadow on the paper behind.

5 Record the angle made by the Sun's shadow at different times of the day.

Garden sundials cast a shadow on a flat plate which has hour lines marked on it.

Sun time

Use your sundial or shadowmeter to tell the time on the next sunny day. Check your results with a watch or clock.

Day and night

Why do we have day and night?

You will need:
- a globe on a stand
- a strong light
- a dark room
- a piece of Blu-tack

1. Find the country where you live on the globe. Mark it with the piece of Blu-tack.

2. Shine the light on the globe, level with the **Equator**.

3. Look at the globe and turn it slowly anticlockwise.

Which countries come into the light at the same time as your own?

Which countries are getting dark when your own country is just getting light?

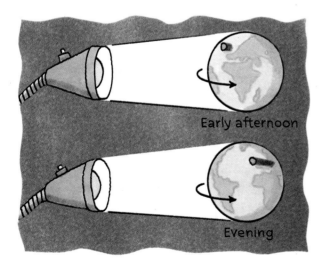

Early afternoon

Evening

Look at the direction of the shadow made by the Blu-tack. It changes at different times of day. The shadow moves like that of the shadowmeter.

Find out:

Look at the globe. Which country would you reach if you could dig through the middle of the Earth?

Use your globe and light to find out if it is morning, afternoon or night there now.

Seasonal day length

You will need:
- a notebook and pencil
- a magnetic compass

1. At sunset, use the compass to find west.
2. In your notebook, sketch the buildings and landmarks that you see.
3. Show the place where the Sun sets.
4. Write down the time and date.

20th December 3.35pm

5. Leave a space in your notebook to do this again at a different time of year.

18th August 8.10pm

Day length

Find the times of sunset and sunrise in the newspaper. What is the length of the day? Do this again one week later. Are the days getting shorter or longer?

Viewed from space, the Earth is half in the light and half in darkness.

Seasons

Why do the seasons happen?

You will need:
- a globe on a stand
- a strong light
- a piece of Blu-tack
- a dark room

1. Shine the light on the globe, level with the Equator. Turn the globe so that the northern part tilts away from the light. This is winter for the northern **hemisphere**.

2. Find the **Arctic Circle**. Slowly turn the globe anticlockwise. Notice that north of this line there is continuous night.

3. Move the globe to the other side of the light so that the northern part tilts towards the light. This is summer for the northern hemisphere. Notice that north of the Arctic Circle there is continuous daylight.

The Earth's axis is tilted by about 23°. The effect of the tilt gives us different seasons. Most globes are set on their stands to show this tilt.

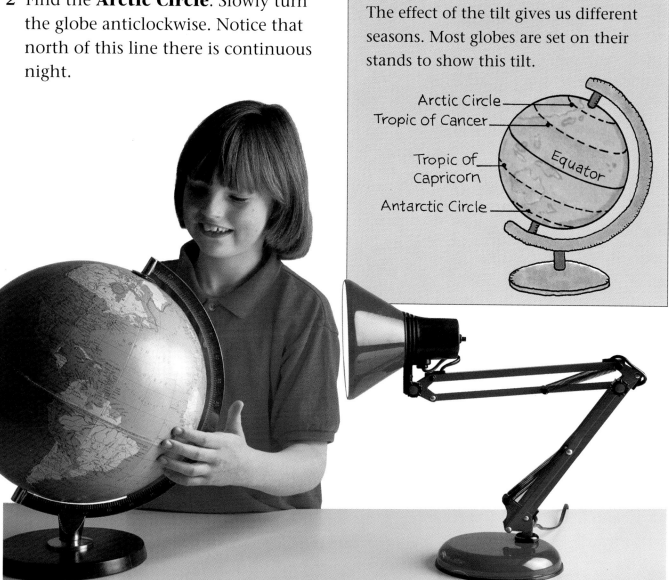

Why is it hotter in summer?

You will need:
- a torch with a narrow beam
- a globe ◆ a dark room

1. Shine the torch straight at the Equator. Look at the ring of light.

2. From the same place, shine the torch at the North Pole. Notice how the beam spreads out.

The same amount of heat is spread over a larger area. This is why it is always hotter near the Equator.

3. Shine the torch on your country in its winter and summer positions. Notice that the beam is more spread out in winter.

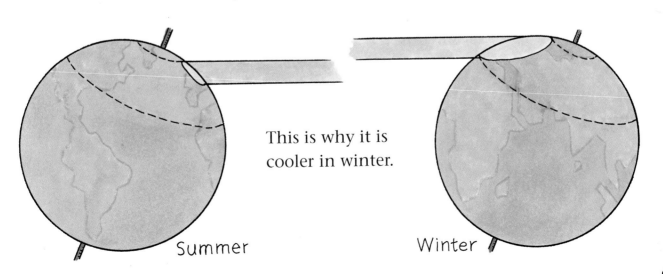

This is why it is cooler in winter.

Summer   Winter

Seasonal changes

Why are the days so short in winter?

You will need:
- a globe on a stand
- a strong light
- a dark room
- a piece of Blu-tack

1. Find your nearest city on the globe and mark it with a piece of Blu-tack. Position the globe so that it is winter in the city.

2. Shine the light on the globe and position it so that the Blu-tack is on the edge of the light. Find out how much daylight the city gets in winter by counting the number of **lines of longitude** which are in the light. (Count around the globe along the city's **line of latitude**).

3. Now move the globe around so that it is summer in the city and repeat the exercise to compare the city's day length in winter and in summer.

4. Find a city a long way north or south of your city which is on the same line of longitude.

5. Which place gets more daylight in winter? Which gets more daylight in summer?

Longitude and latitude

Lines of longitude and latitude are imaginary lines running around the world. Lines of longitude run between the North and South Poles. Lines of latitude run parallel to the Equator. Each line of longitude represents about 40 minutes of time.

Greenwich Meridian

The line of 0° longitude was fixed at Greenwich, Britain in 1884. It was first set to make the same local time for railway trains. Before this, the time of noon could vary within a few kilometres across a country.

Did you know?

In some parts of the Arctic Circle there is continuous daylight for up to four months.

Height of the Sun

You will need:

* a magnetic compass * a notebook and pencil

1. Use a compass to find south.

2. Look south on a sunny day at 3pm. Sketch the buildings and landmarks and draw the position of the Sun. Write down the date and time under your sketch.

This is Moriusaq Village which is in the Arctic Circle. The Sun in the Arctic Circle never sets for certain parts of the year.

3. Leave the page opposite in your book blank so that you can do the same for a different time of year.

What do you notice about the height of the Sun in the two drawings?

3pm December 11th

3pm June 2nd

Solstice

In the northern hemisphere the Sun is at its highest on 21 June. This is the summer solstice. It is at its lowest on 21 December. This is the winter solstice.

Equinox

The 21st March and 23rd September are equinoxes. They are the only two days in the year when the Sun is directly overhead at the Equator. On these two dates, there is equal length of day and night at every point on the globe.

The sky in daylight

How can you safely observe the Sun?

You will need:
- a sunny place ◆ binoculars or a telescope ◆ a tripod
- a large piece of card ◆ a piece of white paper ◆ scissors

Never look directly at the Sun through binoculars or with the naked eye.

1 If you are using binoculars, cover the left eyepiece with a lens cover.

2 Fix the binoculars or telescope on the tripod. Point them in the general direction of the Sun.

3 Cut a hole in the card and fix it around the front of the binoculars or telescope to make some shadow.

4 Hold a piece of paper behind the eyepiece and move it until you see a bright circle of light on the paper.

5 Focus the circle using the focusing ring on the binoculars or telescope.

Can you see darker spots within the bright circle of light? These are sunspots.

Sunspots are cooler areas on the Sun's surface.

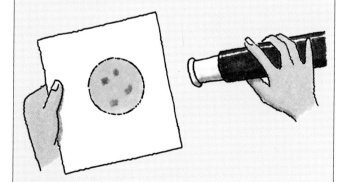

If you were to watch the sunspots over a period of days, you would see them move across the Sun's surface as it rotates.

The sky at night

What is the best way to look at the evening sky?

Binoculars are the best things to use for looking at the Moon and stars. Their strength is shown by two numbers:

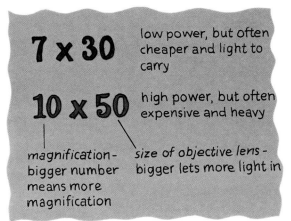

7 × 30 — low power, but often cheaper and light to carry

10 × 50 — high power, but often expensive and heavy

magnification - bigger number means more magnification

size of objective lens - bigger lets more light in

Large telescopes, like this one, are built in clear air at the top of mountains, far away from cities.

Find a place well away from street lights to observe the evening sky.

Adjust your eyes

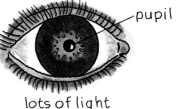

lots of light

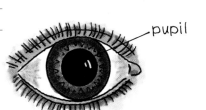

little light

You will need:
- a dark room with a light switch

1 Go into a darkened room.

2 Turn on the light for a minute or two.

3 Turn it off.

4 How long is it before you get used to the dark?

Light gets into your eye through your pupil. Your pupil changes size to adjust to different amounts of light. In a little light your pupil gets bigger. In lots of light your pupil is smaller. It takes time to adjust to seeing in the dark.

Did you know?

Red torchlight won't dazzle you at night. It won't spoil your night vision in the same way as white light.

Be safe

Always tell your parents where you are going and who you are going with.

Never go alone. It's safer and more fun with friends.

Wear warm clothes.

The Moon's surface

You will need:
- a clear, dark sky
- a torch
- binoculars
- a notebook and pencil

1. Study the Moon through your binoculars.
2. Try to find the features shown here.
3. Do a drawing in your notebook. Label all the features you can see.

Ancient rocks

The oldest rocks on the Moon are about 4,600 million years old.

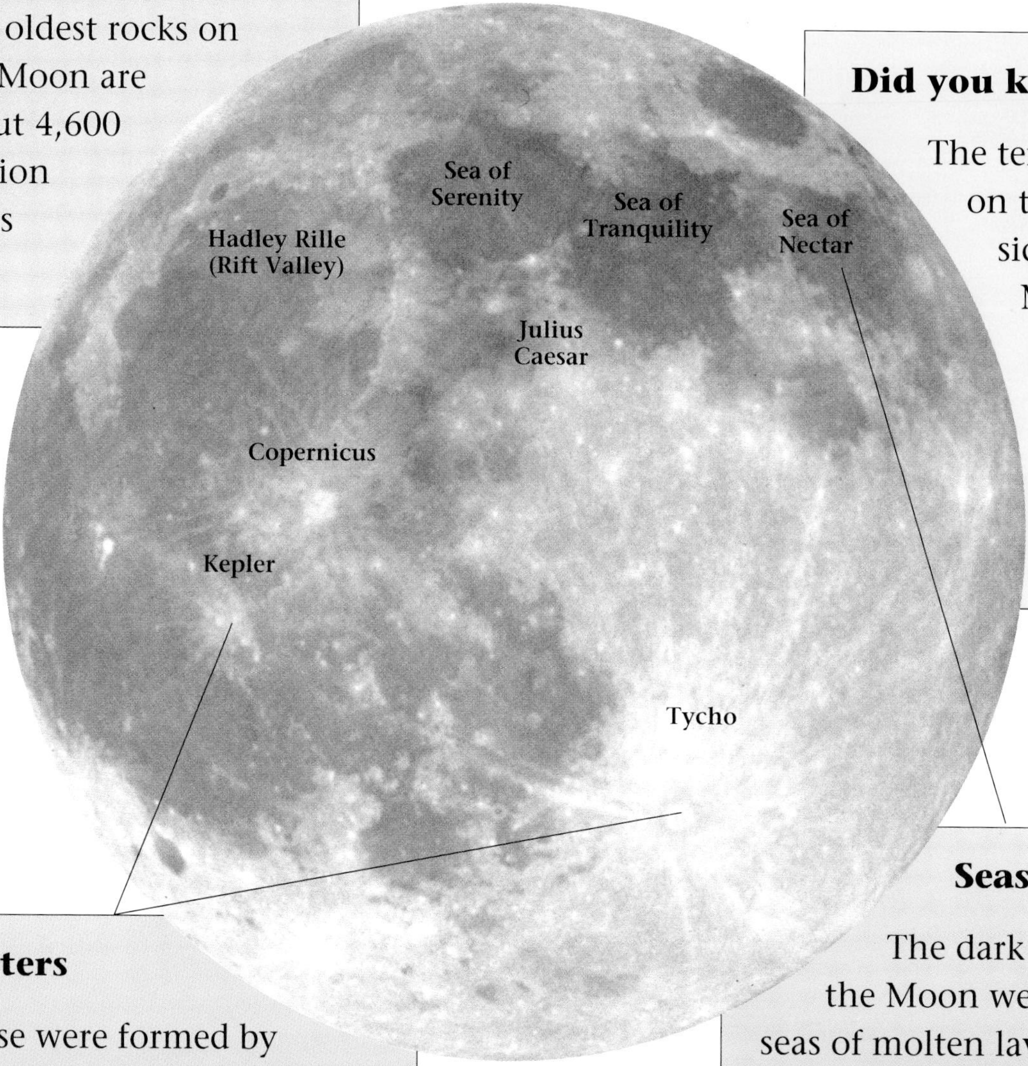

Did you know?

The temperature on the sunny side of the Moon is 100°C. In the shadow it is −150°C.

Labels on Moon image:
- Sea of Serenity
- Sea of Tranquility
- Sea of Nectar
- Hadley Rille (Rift Valley)
- Julius Caesar
- Copernicus
- Kepler
- Tycho

Craters

These were formed by meteorites (lumps of rock) hitting the Moon. Others were formed by long-extinct volcanoes.

Seas

The dark areas of the Moon were once seas of molten lava. They were caused by huge flows of lava from volcanoes. Now they are solid plains of hardened lava.

How were the craters formed?

You will need:
- a small washing-up bowl
- damp sand
- a small, heavy ball or stone

1. Put some sand in the washing-up bowl.
2. Drop the ball or stone on to the sand.
3. Try throwing the ball or stone from close range.
4. Notice the shapes the ball or stone leaves in the sand.

Enormous craters

The biggest craters on the Moon are over 6,000 m deep.

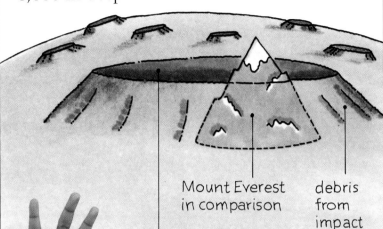

crater walls · Mount Everest in comparison · debris from impact

The Moon's diameter is about the same as the width of Australia.

Papier Mâché Moon

You will need:
- a stiff piece of cardboard
- water
- wallpaper paste
- ripped newspaper
- small card boxes
- milk bottle tops
- silver foil

1. Make a Moon surface using papier mâché.
2. Build your own Moon station using card boxes, milk bottle tops and silver foil.

Phases of the Moon

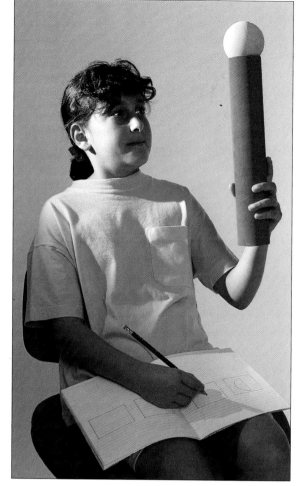

You will need:
- a small, light-coloured ball – to represent the Moon
- a straight beam of light – to represent the Sun
- you – to represent the Earth
- a cardboard tube – to hold the ball
- a stool ◆ a dark room ◆ a notebook and pencil

1. Sit on the stool and balance the ball on top of the tube.

2. Hold the ball up into the light. Draw the shape the light makes on the ball.

3. Turn yourself anticlockwise. Draw the new shape of the light on the ball.

4. Do this for at least four different positions.

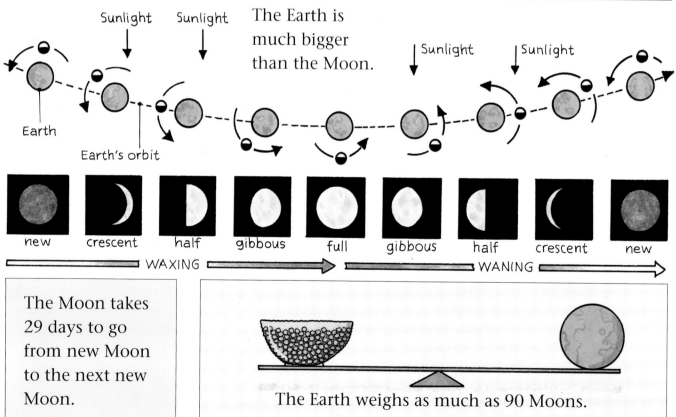

The Earth is much bigger than the Moon.

The Moon takes 29 days to go from new Moon to the next new Moon.

The Earth weighs as much as 90 Moons.

Make a Moonchart

You will need:
- a notebook and pencil
- a ruler
- a piece of A3 paper

1. Draw 30 equal-size squares on a double page of your notebook.
2. Date each square starting with today.
3. Look for the Moon every night for 20 nights.
4. If you see the Moon, draw its shape in the correct square.
5. If the weather is cloudy, put in a cloud symbol.
6. Use the last 10 squares to predict the shape of the Moon for those days. Check to see if your predictions are correct.
7. Make a large Moonchart, like the one in the photo, with your friends, using the results from your Moon watch.

Same face

The same side of the Moon always faces the Earth. We only know about the other side of the Moon because of the spacecraft which have orbited it.

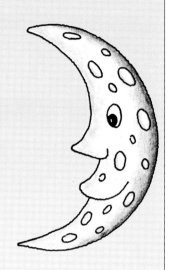

Moonrise and Moonset

You will need:
- a magnetic compass
- a notebook and pencil

1. Use the compass to find east and west.
2. Look east for the place where the Moon rises.
3. Look west for where it sets.
4. Record the times it rises and sets in your notebook.

The Moon is 384,000 km away from Earth. It **orbits** at 1 km per second.

Planet models

What do the planets look like?

You will need:
- a pair of compasses and a pencil
- different-coloured paints
- some card ◆ scissors ◆ a ruler

1. Use the radius shown in the table below to draw circles which represent each planet.

2. Paint each planet according to the colours in the diagram.

Planet	Actual radius (km)	Radius of card model
Mercury	2000	1cm
Venus	6000	4cm
Earth	6000	4cm
Mars	3000	2cm
Jupiter	70000	18cm
Saturn	60000	15cm
Uranus	26000	6cm
Neptune	24000	6cm
Pluto	1500	1cm

Scale

On this scale the Sun would need to be a card disc with a radius of 180 cm.

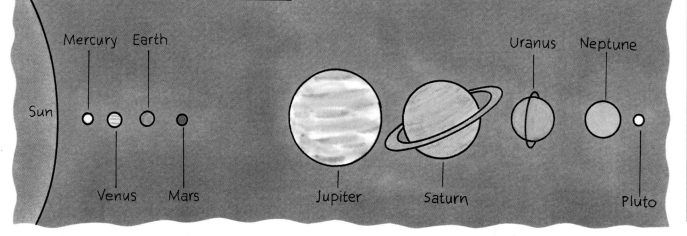

How far apart are the planets?

You will need:

◆ a piece of paper or wallpaper, 1 m long ◆ a ruler ◆ sticky tape ◆ scissors ◆ a felt tip pen

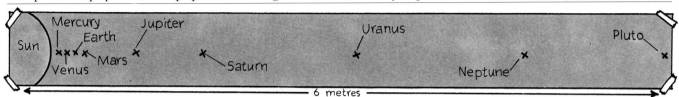

1. Fold the paper in six, lengthways. Cut the paper into six lengths. Stick them together, end to end so that you have a wallchart strip 6 m long. Draw in the Sun at one end.

2. Use this table to mark the positions of the planets on your wallchart with crosses and labels.

Planet	Actual distance from Sun (millions km)	Distance along the wallchart
Mercury	58	6 cm
Venus	108	11 cm
Earth	150	15 cm
Mars	228	23 cm
Jupiter	778	78 cm
Saturn	1427	143 cm
Uranus	2870	287 cm
Neptune	4497	450 cm
Pluto	5900	590 cm

The scale of the wallchart is 1cm: 10,000,000 km.

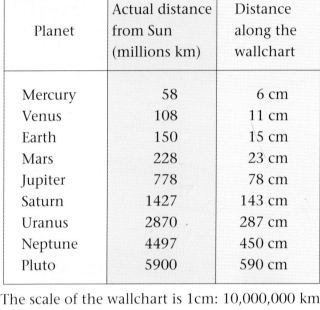

Holiday among the stars

Write a holiday brochure for your favourite planet.

Planet file

Make a set of cards or a computer database for the planets.

Add information to each card as you find it.

Observing planets

You can often see the planets Venus, Mars, Jupiter or Saturn without binoculars. Planets are slightly bigger than the other stars. They are visible at different times of the year.

You can find out which planets are visible now by looking in today's newspaper, or a current astronomy magazine, and making a note of when and where to look for them in the sky.

Planet watch

You will need:

◆ binoculars or a telescope   ◆ a notebook and pencil   ◆ a clear, dark sky

Observing Venus

Venus follows the Sun, so it is easiest to see just after sunset. Venus is often the brightest object in the sky.

1 Draw the position of Venus. Add the date and the time.

2 Use the binoculars to see Venus more clearly.

3 Record its shape in your book.

Observing Mars

Mars is called the red planet. Its soil contains iron oxide (rust), which makes it look slightly red.

1 In your notebook draw the position of Mars and nearby stars.

2 Repeat your observations and records each clear night.

3 Predict where Mars will be in a week's time.

4 Check to see if your prediction was correct.

Venus often has a crescent shape like the Moon.

Observing Jupiter

Jupiter has four large moons visible through binoculars. The moons look like tiny stars in a line near the planet.

1. Draw Jupiter and its moons in your notebook.
2. Repeat the observations on other nights.
3. How do you think the moons are moving?

Write down your ideas.

Observing Saturn

1. In your notebook draw Saturn and any nearby stars.
2. If you have one, use a telescope to look at Saturn and its rings.
3. Draw a picture of your observation in your notebook.

You will need a telescope if you want to see the rings of Saturn.

Venus and Mars are small, rocky planets like the Earth. Jupiter and Saturn are giant planets made of gas.

Did you know?

Saturn's rings are made of chunks of ice and rock which are in orbit round the planet. They may be fragments of a moon pulled apart by Saturn's **gravity**.

Saturn is so light that it would float on water – if you could find a bucket big enough!

Constellations

Constellations are patterns of stars in the night sky. They are like dot-to-dot pictures which make shapes of animals and objects. There are 88 constellations in the sky.

Shoebox constellations

You will need:
- a shoebox ◆ a drawing pin
- scissors ◆ sticky tape
- tracing paper ◆ a pencil
- a constellation picture
(e.g. the Plough from page 23)

1 Place the tracing paper on the constellation picture.

2 Draw pencil dots on the main stars.

3 Place the tracing paper over the end of the shoebox.

4 Use the drawing pin to make holes through the dots and the end of the box.

5 Cut a small viewing hole about 1 cm across at the opposite end of the box.

6 Fix the box lid on with sticky tape.

7 Hold your constellation box up to a light and look through the viewing hole to see the constellation of stars.

Looking for the Plough and Pole Star (northern hemisphere only)

You will need:
- a clear, dark sky
- a notebook and pencil
- a torch
- a magnetic compass

1. Use your compass to find north.
2. In the north, find the seven stars which make a shape like a saucepan with a bent handle. This is called the Plough.
3. Follow the end two stars of the saucepan to Polaris, the North Pole Star.

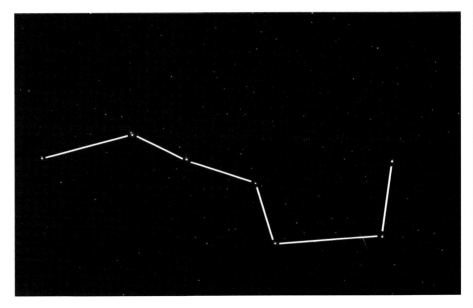

This photo shows the Plough constellation. You cannot see the Pole Star in this photo, but the diagram below shows you how to find it.

Did you know?

The Pole Star is almost directly above the Earth's North Pole. It is the only star which stays in the same place as our Earth turns on its axis.

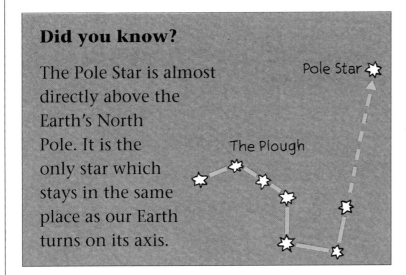

Southern hemisphere

In the southern hemisphere, the sky is different from the north. You can see most of the constellations at the same time as you would see them in the north. However they appear in different positions, often upside-down compared to the northern sky.

The Southern Cross is a diamond-shaped constellation which you can only see in the southern hemisphere.

Observing the night sky

The sky in March, April and May

As the Earth orbits the Sun, we see different stars at different times of year.

You will need:
- a torch ◆ a notebook and pencil
- binoculars or a telescope ◆ a magnetic compass

Remember that if you live in the southern hemisphere, the constellations may be a different way up from those in the pictures here. However, you can find all of them by looking north instead of south.

Leo the Lion

1. Find the stars of Leo in the south (look for the big backward question mark).

2. Draw the stars of Leo in your notebook.

3. Can you sketch the shape of a lion around the stars?

4. Find Regulus, the bright star at the bottom of the question mark.

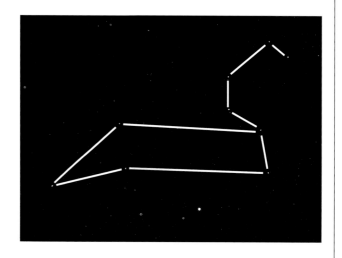

Bootes the Herdsman

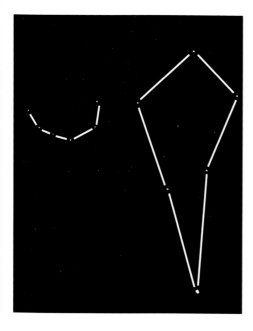

1. Find the Plough almost overhead.

2. Follow the handle of the Plough down to a bright orange star called Arcturus.

3. Look above Arcturus for stars in the shape of a kite. This is Bootes. (In the southern hemisphere you can spot Bootes by looking north in March for the shape of a kite.)

4. Near the top left of the kite is a curve of 6 stars. It is the Northern Crown.

5. Draw Bootes and the Crown in your notebook.

6. Look at Regulus and Arcturus with binoculars. Record their colours in your notebook.

The sky in June, July and August

The Great Triangle

1. Look south for 3 bright stars making a huge triangle. These are Deneb (top), Vega (right) and Altair (bottom).

2. Look at the three stars through binoculars. Record their colours in your notebook.

3. Look below Deneb for 3 stars which, together with Deneb, make a cross. This is Cygnus the Swan.

4. In your notebook, make your own drawing of the stars in the Great Triangle. Label the stars and constellations.

5. Use binoculars to look at Cygnus. You will see thousands of stars. This is part of the band of stars that make up our **galaxy**, called the **Milky Way**.

Did you know?

The colours of stars depend on their temperatures. The hottest stars are blue, the least hot are red.

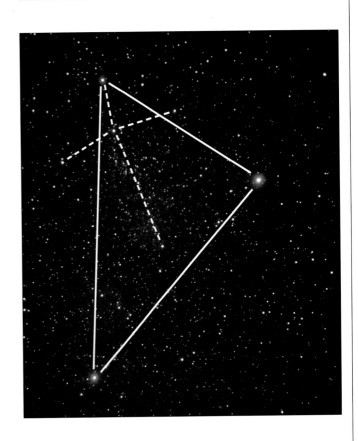

Did you know?

Our galaxy is like a giant frisbee. When we look into our galaxy we see a band of stars stretching across the sky. This is the Milky Way.

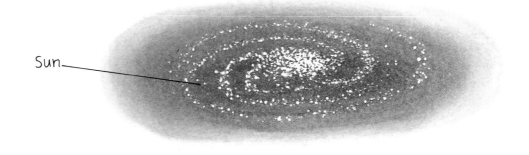

Observing the night sky

The sky in September, October and November

Pegasus and Andromeda

1. Look south for four stars making the huge Square of Pegasus.

2. Draw the stars of Pegasus in your book.

3. Join the stars to show Pegasus as the flying horse (he is upside-down).

4. Look for a line of stars running from the top left of the square. This is Andromeda, the princess.

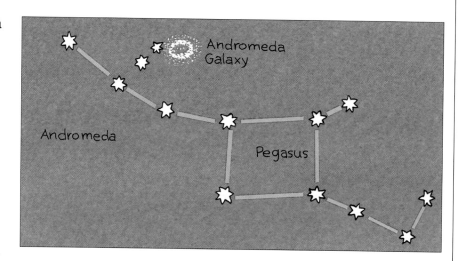

5. Look carefully above Andromeda's second star. You will see a faint, fuzzy glow – this is the great galaxy Andromeda.

Andromeda is named after a character from ancient Greek mythology. Andromeda was the daughter of Queen Cassiopeia. When a sea monster threatened the coast of their country, Andromeda was chosen as a sacrifice. She was rescued just in time by Perseus.

Did you know?

The galaxy Andromeda is the closest spiral galaxy to our own Milky Way. Even so, its light takes over two million years to reach us.

The sky in December, January and February

Orion the Hunter

1. Look in the south for the stars of Orion. You can find them from the 3 stars in a line which make the hunter's belt.

2. Draw the stars in your book. Label the stars Betelgeuse (top left) and Rigel (bottom right).

3. Note the colours of Betelgeuse and Rigel.

4. Look below the 3 stars in the belt – imagine a sword hanging down. You will see a hazy patch of light. This is the Orion Nebula, a huge cloud of glowing gas.

5. Look at the nebula through binoculars or a telescope.

> **Did you know?**
>
> Betelgeuse is a red supergiant star. If it replaced the Sun it would stretch out beyond the orbit of Mars.

Taurus the Bull

1. Follow the belt of Orion upwards to a bright orange star. This is Aldebaran – the eye of Taurus the Bull.

2. Find two bright stars above and to the left of Aldebaran. These are the bull's horns. (See left).

3. Look above and to the right of Aldebaran for a twinkling patch of stars. This is the Pleiades star cluster, or Seven Sisters.

4. Look at the Pleiades cluster through binoculars. See if you can count the stars in the cluster.

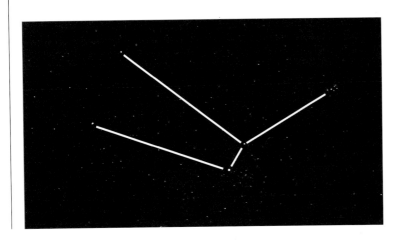

Meteors

Fragments of rock and dust travelling through space are called **meteors**. When meteors enter the Earth's atmosphere, they are travelling at enormous speed. Friction makes the material burn up. It leaves a streak of light which people call 'shooting stars'. Meteors often fall in numbers. This is called a meteor shower.

When to look for meteors

Look at the meteor shower table below and plan a good night for a meteor watch with your friends.

A meteor, or 'shooting star', plunging through space, leaving a fiery trail.

Name of Meteor Shower	Best time to look for them	Maximum number per hour
Quadrantids	January 4th	100
Lyrids	April 21st-23rd	10
Aquarids	May 4th-6th	20
Aquarids	July 27th	30
Perseids	August 9th-14th	70
Orionids	October 19th-21st	30
Taurids	November 7th	10
Geminids	December 10th-15th	60

Did you know?

Meteor showers come from pieces of rock and dust left by old comets.

Recording a meteor shower

You will need:
- a clear, dark sky
- a comfortable chair
- a piece of A3 paper
- a torch
- coloured pencils
- a magnetic compass

1. Use your compass to find east and sit facing that direction.

2. Draw an outline of the horizon in the east. Add in the bright stars that you see.

3. When you see a meteor, draw its track on your diagram, using an arrow to show the length and direction.

4. Beside the arrow write down the meteor's:

- brightness (bright, medium or dim)
- colour
- whether it left a 'train' (a glow behind it)
- unusual features eg: explosions along its path

What do you notice about your meteor tracks?

Meteor showers come from the same point in the sky. This point is called the meteor radiant.

East

True or false?

1. The Earth rotates anticlockwise.
2. The line of 0° longitude passes through Greenwich (UK).
3. You could swim in the Sea of Tranquility.
4. We always see the same face of the Moon.
5. The biggest planet is Neptune.
6. Polaris is another name for the Pole Star.
7. Saturn is the only planet with rings.
8. It takes over two million years for light to reach us from the galaxy Andromeda.

All the answers are in this book.

Glossary

Arctic Circle The line of 70° (N) latitude on the globe. On and to the north of this line on 21 June there is continuous daylight.

Axis (Earth's) The Earth spins round its axis, tilted at an angle of 23°.

Binoculars A form of double telescope with an eyepiece for each eye.

Constellations Groups of stars which appear in the same part of the sky.

Equator The line around a ball which is at right angles to the axis.

Galaxy A huge collection of stars separated by enormous distances. Galaxies contain billions of stars. There are billions of galaxies.

Gravity The force of attraction which a very large mass, like a planet or a Moon, exerts on objects. The pull of the Sun's gravity keeps all the planets in orbit.

Hemisphere The Earth is split up by the Equator into two half spheres, the northern and southern hemispheres.

Lines of latitude Imaginary lines that run around the globe parallel to the equator.

Lines of longitude Imaginary lines that run around the globe from the North to the South Pole. They are used to represent time differences around the world.

Meteors Pieces of rock dust burning up in the Earth's atmosphere.

Milky Way This is the galaxy which we are part of. It contains 100 billion stars. We see the Milky Way as a band of stars across the night sky.

Orbiting To circle another object. The Earth orbits the Sun. The Moon orbits the Earth.

Planet A ball of rock or gas which orbits a star. There are nine planets in the Solar System. At the time of writing no planets have been observed orbiting any other star.

Pole, North and South The two places on the surface of the Earth which are furthest from the Equator.

Seasons The predictable changes in the climate that happen during part of the year. Seasonal change is most obvious in the northern and southern parts of the Earth.

Tropic of Cancer The line of latitude which is 23° north of the Equator. On 21 June the Sun is directly overhead the tropic at noon.

Tropic of Capricorn The line of latitude which is 23° south of the Equator. On 21 December the Sun is directly overhead the tropic at noon.

Waning Getting smaller. The Moon wanes as it goes from full Moon to new Moon.

Waxing Getting bigger. The Moon waxes as it goes from new Moon to full Moon.

Books to read

An Early Start to Earth and Space by Roy Richards (Simon & Schuster, 1992)
Be an Expert Astronomer by Ian Graham (Watts, 1991)
Gems Guide to The Night Sky by Ian Ridpath (Collins Gem Guide, 1994)
Teaching and Understanding Science by Graham Peacock and Robin Smith (Hodder Headline, 1992)

Monthly magazines include:
Astronomy, Sky and Telescope, Astronomy Now

For more information, contact:

Junior Astronomical Society, 36 Fairway, Keyworth, Notts NG12 5DU, England

The Royal Astronomical Society of Canada, Dept. M, 136 Dupont St, Toronto, Ontario M5T 1V2, Canada

Queensland Astronomical Society, PO Box 101, St. Lucia 4067, Queensland, Australia

Notes for parents and teachers

Pages 4–5 It is less confusing when describing the rotation of the Earth if you always imagine that you are looking down on the North Pole. In that case the instruction anticlockwise is easy to understand. Sundials are not accurate to the minute. Many sophisticated sundials include a table which helps to make them more accurate. The speed of rotation at the surface of the Earth decreases the further north or south you go from the equator. This gives rise to the Coriolis effect which bends weather systems.

Pages 6–7 Make the piece of Blu-tack as raised as possible but make sure it fits under the globe's support. When calculating daylength, try to avoid the periods around the summer and winter solstices as you may get confusing results. Of course it is impossible to dig through the Earth. In fact no one has even drilled through the relatively thin crust to the mantle layer.

Pages 8–9 The seasons occur because of the tilt of the Earth. If it wasn't for this tilt there would be no seasons. The tilt of the Earth remains constant. It is the effect of moving to the other side of the Sun which causes the seasonal effect. Shadows are longer in the winter because the Sun is lower in the sky.

Pages 10–11 Pairs of cities which are on the same line of longitude include London in England and Valencia in Spain. Cairns and Sydney in Australia and Miami and Toronto in North America are also good contrasts. On the Arctic and Antarctic circles there is only one day each year when there is 24 hours of daylight. Further north or south of these lines respectively, the number is greater.

Pages 12–13 In general binoculars are a better buy than a telescope. Buy the best you can afford but good results can be obtained from an inexpensive pair. It is most important to find a place well away from street lights.

Pages 14–15 The Moon's volcanoes died out about 3 billion years ago. Very few rocks of this age are left on the Earth's surface because of continued geological and weathering activity. The gravity on the Moon's surface is one-sixth that of the Earth's surface.

Pages 16–17 The Moon orbits the Earth anticlockwise. When drawing the phases of the Moon it is important to remember to draw them as seen from the Earth. One tip is to say whether the left-hand or the right-hand part of the Moon is lit up.

Pages 18–19 It is impossible to devise a practical scale which shows any detail on the planets and also shows them the correct distance apart. When you use a scale which shows the Sun as a beachball and Jupiter as a pea, Jupiter has to be placed hundreds of metres from the ball. There is a lot of space between the planets!

Pages 20–21 The exact positions of the planets can be found in newspapers and monthly astronomy magazines. If using binoculars, lean against something firm, like a wall or fence, to hold them steady. Galileo was the first to observe the four moons of Jupiter; this led him to understand that everything did not go round the Earth as the church of his time taught.

Pages 22–23 The Plough is always visible in the night sky from the northern hemisphere. It can be in different positions around the Pole Star at different times of year.

Pages 24–25 When going out to observe constellations, leave at least 15 minutes for your eyes to adjust. You will be astonished how much more you can see. Choose an observing site away from any strong lights. The three stars of the Great Triangle are very bright and obvious. The other stars such as Sagitta are much dimmer. The Milky Way might be difficult to see from a city with street lights.

Pages 26–27 The Andromeda galaxy is visible with the naked eye but it is much clearer with binoculars. Constellations are often named after gods and animals from ancient Greek mythology.

Pages 28–29 Meteor trails can be of different colours, ranging from white to yellow and green. Meteor showers usually come from constellations which rise in the eastern sky in early evening. As the constellations move across the sky, the meteors' source will move with them.

Index

Aldebaran 27
Altair 25
Andromeda Galaxy 26
Arctic Circle 8, 11
Arcturus 24
axis, Earth's 4, 8, 23

Betelgeuse 27
binoculars 12–14
Bootes the Herdsman 24

comets 28
constellations 22–27
craters 14–p15
Cygnus 25

day, and night 6–7
daylight 8, 10–13
Deneb 25

Earth 4, 16, 19
Equator 9
equinoxes 11

galaxies
  Milky Way 25
  Andromeda 26
Great Triangle 25

Jupiter 19–21

Leo the Lion 24

light 12–13
lines of latitude 10
lines of longitude 10

Mars 19–21, 27
Meridian, Greenwich 10
Mercury 19
meteorites 14
meteors 28–29
midnight sun 11
Milky Way Galaxy 25–26
Moon 13–17

Nebula, Orion 27
Neptune 19
Night 13
Northern Crown 24
northern hemisphere 8, 11
North Pole 4, 10, 23

Orion 27

Pegasus 26
Perseus 26
planets 4, 18–21
Pleiades star cluster 27
Plough 22–24
Pluto 19
Polaris (North Pole Star) 23

Regulus 24
Rigel 27

rotation, Earth's 4

Sagitta 25
Saturn 19–21
seasons 4, 7–11
shadowmeter 5
solstice, summer and
  winter 11
Southern Cross 23
South Pole 10
southern hemisphere 23
stars, colours 25
summer 9–10
sun 4, 12, 18–19
sundials 4–5
sunset and sunrise 7
sun, height 11
sunspots 12

Taurus the Bull 27
telescope 12–13
time 4–6, 10

Uranus 19

Vega 25
Venus 19–21

winter 8–10